NONE
DARE
CALL
IT
TREASON!
BOOK 9

Inexcusably
Arming America's
Enemies!

Robert W. Pelton
$4.95

"Treason doth never prosper,

"What's the reason?

"Why if it prosper,

"None dare call it treason."

John Harrington

Printed in America
On Recycled Paper
In
Charleston, South Carolina

Published in America
By
The Freedom & Liberty
Foundation Press
Knoxville, Tennessee

Dedicated
To

The greatest, most generous, most benevolent and most powerful nation on the face of the earth – and the only country in the history of the world to have been founded on Biblical principles.

A nation can survive its fools, and even the ambitious. But it cannot survive treason from within.

An enemy at the gates is less formidable, for he is known and he carries his banners openly.

The traitor moves among those within the gates freely, his sly whispers rustling through the galleys, heard in the very hall of government itself.

For the traitor appears not traitor. He speaks in the accent familiar to his victims, and he wears their face and their garments, and he appeals to the baseness that lies deep in the hearts of all men.

He rots the soul of a nation - he works secretly and unknown in the night to undermine the pillars of a city - he infects the body politic so that it can no longer resist.

A murderer is less to be feared.

Cicero, 42 B.C.

CONTENTS

Forward

Independence Hall Where the Declaration of Independence Was Signed.

Our glorious Declaration of Independence is a timeless divinely inspired masterpiece given to mankind through the anointed pen of Thomas Jefferson.

The grand and unmatched United States Constitution is indisputably the product of Providential guidance and wisdom and certainly not a document which evokes whimsical interpretations with the changing political climates.

All Americans have a moral obligation to stand up and be counted in these trying times!

Abraham Lincoln boldly declared: *"To sin by silence when they should protest, makes cowards of men."*

William Lloyd Garrison capsulized it best: *"As a free man who is determined to remain free -- I do not wish to think or speak, or write with moderation. "Tell a man whose house is on fire to give a moderate alarm; tell him to moderately rescue his wife from the hands of a ravisher; tell the mother to gradually extricate her babe from the fire into which it has fallen -- but urge me not to use moderation in a course like the present."*

Senator Barry Goldwater, 1964 Presidential candidate was castigated and verbally crucified by the media.

He simply stated this simple truism: *"Extremism in the pursuit of Liberty is no vice."*

This good and moral man of character soundly rocked the boat of the propagandists. He was as a result soundly defeated in the election.

The alarmed media wolves panicked the voters with their jeers and sneers and insane howls about this man's lack of *"moderation!"*

It can honestly be said that through the Providential genius of our Founding Fathers, the remaining remnants of the original American Constitutional Republic still provides more freedom, opportunity and abundance for mankind than is found in any other nation in the world.

This is true despite decade after decade of unabated treason and treachery promulgated by innumerable traitorous individuals found buried in the twiddle dee – twiddle dum administrations of both the Democrats and the Republicans.

An informed and active, not a media brainwashed electorate, is the only antidote to further prostitution of, and the ultimate destruction of, what Benjamin Franklin called our Republic.

Preface

"Treason against the United States shall consist only in levying war against them, or in adhering to their enemies, giving them aid and comfort."

U.S. Constitution. Article 111, Section 3

What is your treason I.Q.?

If you can answer the following questions, it's high.

If you miss one or more, you should read the *None Dare Call It Treason* series!

Who was behind allowing Red Chinese soldiers take airborne training at Fort Benning, Georgia?

Is this not treason?

Why was South Vietnam, South Africa, Rhodesia and numerous other American friends deliberately betrayed to the forces of evil?

Is this not treason?

Why was our friend Chiang Kai Shek not so gently coerced into a Communist dictatorship by highly placed subversives in the State Department?

Is this not treason?

Why was Cuba treasonously delivered into the clutches of Communist revolutionary Fidel Castro?

Is this not treason?

Why have untold millions of dollars consistently been used to prop up faltering Red dictatorships and to assist Communist terrorists in overthrowing non-Communist governments?

Is this not treason?

What American company sold nuclear reactors to Communist Occupied Romania?

Is this not treason?

Name the company that provided Communist Hungary with a factory designed to make 1.5 million light bulbs daily?

Is this not treason?

What well known oil company invested $1 billion for oil exploration in Communist Occupied Angola?

Is this not treason?

Can you name the American company who treasonously built and equipped a $10 million electronics plant near Warsaw for the Polish slave labor tyranny?

Is this not treason?

These are questions to which every American should rightfully have an honest answer.

Unfortunately most do not!

Tragedy was carefully orchestrated by traitors in our Government and the media with regard to Cuba, Vietnam, Laos, Cambodia, Rhodesia, China, El Salvador, Nicaragua and many other countries.   Anastasio   Somoza was the former President of free Nicaragua.

He offered this startling insight in his 1980 book, *Nicaragua Betrayed*: *"I have factual evidence that the betrayal of Nicaragua was not perpetrated out of ignorance, but rather by design."*

Somoza was soon after assassinated!

Is this not treason?

John Lehman, Secretary of the Navy, made this shocking statement on May 25 to the 1983 Annapolis graduating class: *"Within weeks many of you will be looking across just hundreds of feet of water at some of the most modern technology ever invented in America.*

"Unfortunately, it is on Soviet ships."

Is this not treason?

Earl E.T. Smith was the American Ambassador to

Cuba when it was similarly delivered to the Communists.

He makes this concise comment on July 14, 1986: *"Nicaragua is Cuba all over again."*

Can you name the company that paid the Communist dictatorship in Angola over $600 million annually in taxes and oil royalties.

This money bought new Soviet jets, tanks and helicopter gunships.

And it paid Castro for supplying 35,000 imported Cuban mercenaries who keep the Angolan people enslaved.

Is this not treason?

Stressed retired Brigadier General Andrew J. Gatsis on August 11, 1986: *"Though aware of the Communist goal of world domination, the average U.S. Citizen refuses to believe that the real threat comes from governmental officials and their non-governmental confederates who secretly espouse the same objectives as the openly avowed Communists."*

Anthony Sutton stated in his 1986 book *The Best Enemy Money Can Buy:* *"We now have the formidable task of bringing these gentlemen to the bar of justice to publicly answer for their private and concealed actions."*

The *None Dare Call It Treason* series certainly won't win accolades from the United Nations or the State Department!

Nor will Harvard feel compelled to bestow an honorary degree upon the author!

Harvard Law School was the spawning ground for an incredible number of Red agents. Included were members of the first Soviet spy ring ever to be exposed in our government.

Reed Irvine aptly commented in July of 1986: *"Indeed, it has long been a joke among refugees from Eastern Europe that there are more Marxists at Harvard than there are in the Soviet Union, or Poland, or whatever Communist country the refugee called home."*

23

 The Honorable Ezra Taft Benson said: *"The truth must be told even at the risk of destroying, in large measure, the influence of men who are widely respected and loved by the American people.*

"The stakes are high. Freedom and survival is the issue."

Treason is still a most serious federal offense.

The *None Dare Call It Treason* series examines the reasons for and the Americans behind the fall of freedom and the rise of tyranny throughout the world!

Has anything really changed?
You Decide!

Treason

Whoever, owing allegiance to the United States, levies war against them or adheres to their enemies, giving them aid and comfort within the United States or elsewhere, is guilty of treason and shall suffer death, or be imprisoned not less than five years and fined not less than $10,000; and shall be incapable of holding any office under the United states.

U.S. Code, Title 18, Section 2381

Whoever, owing allegiance to the United States and having knowledge of the commission of any treason against them, conceals and does not, as soon as may be, disclose and make known the same to the President or to some judge of the United States, or to the Governor or to some judge or justice of a particular state, is guilty of misprision of treason, and shall be fined not more than $1000 or imprisoned not more than 7 years or both.

U.S. Code, Title 18, Section 2382

26

The Inexcusable Arming of America's Enemies

Treason: *"Betraying one's government… revealing military secrets to the enemy."*
Webster's Elementary Dictionary

Former Congressman John Breckinridge talked to Soviet First Deputy Minister of Defense N.V. Ogarkov while visiting Communist Occupied Russia in 1978.

He was warned: *"The United States has always been in a position where it could not be threatened by foreign powers.*

"That is no longer true.

"Today the Soviet Union has military superiority over the United States and henceforth the United States will be threatened.

\ *"You had better get used to it!"*

It's an acknowledged fact that Communist Occupied Russia is terribly inept when it comes to advances in science and technology.

The backward Soviets couldn't possibly have become a major military power without the help of the United States.

No one really believes that a dictatorship incapable of even manufacturing a common light bulb could possibly create and build its

own sophisticated missiles and high-tech weaponry!

No one really believes that a dictatorship even incapable of producing an automobile could possibly create and build its own tanks, assault helicopters, and jet aircraft!

America spends billions on defense in order to offset the USSR's advanced war-making capabilities.

Ironically the Soviets would possess little if any of this without U.S. technology transfers and arms sales.

There's absolutely no question about the correlation between American assistance and the military might of the Evil Red Empire.

There is ample reason to believe that the Soviet Union as well as all of the slave bloc nations and Communist Occupied China were deliberately armed by the United States.

American aid, trade, money and business dealings are the only things responsible for the Red dictatorship's ability to flex its muscles!

All of this is well known by the highest ranking leaders in Washington!

Yet, these same officials go to phenomenal lengths to hide these facts from the American people.

All Americans should heed the words of retired Brigadier General Robert L. Scott, Jr.: *"Our threat is not from abroad. The danger is internal, and the solution can only be internal.*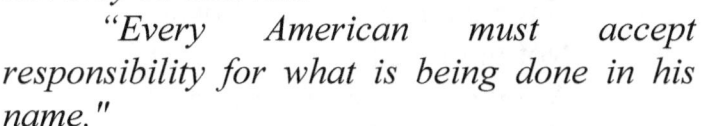

"Every American must accept responsibility for what is being done in his name."

As Cicero once wisely advised: *"A nation can survive its fools, even the ambitious, but it cannot survive treason from within."*

The Russians were promoted as the *"good guys"* when the U.S. began treasonously arming them many years ago.

This was done under the transparent guise that there was a serious Chinese-Soviet split.

If the U.S. assisted the Soviets Americans were told, the geriatric despots who run the Red dictatorship would surely mellow.

In case of war with the Communist Chinese *"bad guys,"* the Soviet *"good guys"* would be America's allies.

So the United States injudiciously sold, traded and gave the insatiable Russian Bear incredible amounts of weaponry and computers and factories.

The Soviet Union became a first rate Made-In-The-USA military power.

Have the Russians mellowed?

Not in the least!

Are they America's friend?

Certainly not!

Every Soviet dictator up to and including suave American media favorite Mikhail Sergeyvich Gorbachev has been unswerving in his commitment to one day conquer the United States!

U.S. News & World Report observed: *"Russia's leaders are shaping a military force as if they expect World War III to erupt at dawn tomorrow."*

In 1930, Ford Motor Company contracted to build Communist Occupied

Russia a state-of-the-art mass-production vehicle factory.

Ford and the Austin Company designed, advised, gave technical assistance and furnished the equipment for a plant in Gorki, 250 miles east of Moscow.

Gorki is a closed city where important Russians who get out of line are banished into *"internal exile"* by Communist Party bosses.

The Gorki plant was one of the largest in the world.

It was designed to annually produce 140,000 cars, trucks, and buses.

The factory was constructed with slave labor under the supervision of Albert Kahn, the builder of Ford's River Rouge complex near Detroit.

Ford specialists taught Russian forced-laborers the finer aspects of assembling mass-produced vehicles.

Soviet engineers were even brought to America for training at the River Rouge plant.

Yes, it can be truthfully said that Henry Ford established the Russian dictatorship's auto, truck and bus industry!

And Gorki was heralded as Russia's version of Detroit!

Nevertheless, car production in today's Soviet Union is negligible.

More autos are stolen annually in the United States than are produced in the USSR.

There are fewer privately owned automobiles in Communist Occupied Russia than there are cars owned by South African blacks.

Moreover, the average Russian citizen has little need for an automobile.

The entire country contains fewer paved roads and highways than does Massachusetts.

But these trivia tidbits are beside the point!

The Soviets did not use the Ford plant to mass-produce automobiles, commercial trucks or buses!

Instead, the assembly lines spewed forth untold numbers of Red Army vehicles.

1930 was also the year Detroit's Arthur J. Brandt Company built their big ZIL factory for the Russians.

C.P. Weeks was the Vice president of Hercules Motor Corporation in Canton, Ohio.

He declared the factory was: *"By far the largest and best equipped plant in the world devoted solely to the manufacture of trucks and busses."*

Not surprisingly the Brandt factory didn't make commercial trucks or busses.

It produced *armored trucks* and *howitzer tractors.*

The William Sellers Company of Philadelphia received a contract from Communist Occupied Russia in 1938.

They supplied Stalin's murderous dictatorship with the heavy machinery required to produce 12-inch thick armor plate.

In March 1939, Electric Boat Company of Groton, Connecticut, was selected to supply the blueprints, specifications and construction services necessary to build the  backward Soviet Union's first submarine.

Red moles, other traitorous leftists, and immoral opportunists buried in the bureaucratic honeycombs of the government have for years been illegally making all the arrangements for the arming of Russia!

And for all the other Red dictatorships around the world as well.

These criminal Marxist dictatorships have been shipped a never-ending supply of strategic war-related goods.

For example, take a look at the heavily Red infiltrated *(Hiss, White, Abt, Pressman, etc.)* Democratic Administrations of Roosevelt and Truman.

These Red moles were instrumental in sending between $11 and $12 billion in weaponry and other goodies to their Soviet comrades between 1942 and 1946.

Yet no one was ever arrested and charged with being a traitor!

Why not?

No one was ever prosecuted and sent to prison for treason!

Why not?

The array of military items given to the Russian enemy was absolutely phenomenal.

It pretty much covered anything and everything necessary for the production of military goods.

Included were 219,000 tons of scarce copper cable and wire.

More than 1,300 diesel engines for ships.

More than 400 destroyers and other combat ships.

Vast quantities of special machine tools and industrial machinery.

Well over 300,000 tons of explosives.

More than 14,000 planes.

Nearly 500,000 jeeps, tanks and trucks.

Finally in 1951 the *Battle Act* was passed by Congress.

This was a desperate effort to close the arms giveaway flood gates.

A concise list of non-tradable military-related items was incorporated in the legislation.

Harold E. Stassen became Eisenhower's administrator of the *Battle Act*.

He took it upon himself to drastically revise the list of non-tradable items downward in 1954 to benefit America's many Communist enemies.

The end result was a booming increase in strategic trade between the United States and the entire Soviet Bloc.

The Communist war machine was now allowed to freely obtain machine tools, generators and other highly strategic goods.

These included electronic equipment used on guided missiles and in the field of atomic energy.

Many items of military importance yesterday per Stassen's illogic were somehow overnight no longer considered *"strategic."*

As a direct result of these inexcusable pro-Red arming activities the military might of the Soviet Union was dramatically expanded as was fully explained in *East West Trade*, a 1969 *Senate Report*.

Nevertheless, in 1956, Congress was warned not to stop military aid to America's *"friend"* who made no secret of their infamous Gulag Archipelago.

This atrocity consists of well over 2,000 known slave labor concentration camps!

They're always filled with at least 15 million prisoners and sometimes with as many as 30 million!

Better than 40 percent of Soviet industry is wholly dependent upon slave labor.

"These victims—men like you and me work out the days or years until they are no longer worth famine-rations to their captors," explained the late Robert Welch. *"As we sit in our warm homes, after a happy meal with our*

families, and turn on our television sets or radios, it's hard for us to think of a man just like ourselves always half-starved, always half-frozen, haggard and hopeless, remembering the days when he too was free, as he is brutally driven to finish up the literal exhaustion of his body in labor for the benefit of the very tyrant who has enslaved him."

During the years Stalin ruthlessly ruled the Russian dictatorship the annual prisoner replacement rate exceeded an astounding 50 percent!

More than half the prisoners died of malnutrition, torture, freezing, overwork and murder.

Nothing changed in this regard in Mikhail Sergeyvich Gorbachev's despicable Soviet dictatorship!

In his 1948 book, *Forced Labor in the Soviet Union*, Professor David Dallin revealed what had happened to a shipment of slave laborers on the way to Siberia: *"The Dzhurma sailed from Vladivostok to Ambarchik (over 4,00 miles) carrying about 12,000 prisoners.*

"The ship reached the Arctic Ocean too late in the season, and was caught in packed ice near Wrangel Island.

"The Dzhurma, when it finally arrived in Ambarchik did not land a single prisoner.

"However, what mattered for the government was not the loss of prisoners but the fact that the valuable ship was saved."

"The Soviet government uses prisons and labor camps to crush those who oppose official ideology and policies," charged Anatoly T. Marchenko. *"The Soviet government views this as its sovereign right and a purely internal affair."*

Much of the Soviet Union's slave labor is not Russian by citizenship.

For example Communist Occupied Russia and Nazi Germany jointly invaded Poland in 1939.

A million-and-a-half Poles were deported to slave-labor camps in Central Asia and Siberia!

The same thing happened when the hungry Russian Bear gobbled up Latvia, Estonia, and Lithuania.

Said Michael F. Connors: *"Of the approximately three million inhabitants of Lithuania in 1939 fully 500,000 ended up in Siberian slave-labor camps!*

This was done in pursuance of a policy of systematic depopulation as a prelude to colonization by more 'reliable' Russians."

Cronid Lubarsky was an astronomer who spent five years in Russian labor camps and prisons.

He revealed: *"Essentially, there is not a significant area of the Soviet economy in which prison labor is not exploited.*

"Metals processing, the chemical industry, the manufacture of clothing and of machinery, agriculture, mining -- forced labor is used in all of them."

David Satter of the *Wall Street Journal* agreed: *"Forced labor is by no means exceptional in the Soviet Union.*

"It is an integral part of the economic system, and it is extremely doubtful whether the current Soviet economy could function without it."

Former gulag resident Georgy Davydov gave testimony before Congress in 1983 where he described life in a slave labor camp: *"Using Regular Women's Camp UTs-267/10 as an example (in Gornoye in Maritime Kray) there are 2,000 women in a camp designed for 500.*

"Water is brought in from the outside and is therefore in short supply.

"Baths are rare.

"The laundry has only 20 tubs for 2,000 women!

"There are only two paramedics (and no physician) in the medical unit where the possession of medicine by inmates is prohibited.

"Fungus infections, dysentery, and jaundice are rife in the camp."

Julia Voznesenskaya was arrested by the KGB in Leningrad and tried for *"anti-Soviet slander."*

She was convicted and quickly sent to a Siberian slave labor camp.

Said Joseph Harriss in *Readers Digest* of September, 1983: *"Shifts stretched to 12 hours to meet impossibly high production quotas.*

"Those who failed to meet them had their meager food ration cut.

"With gallows humor, she and her fellow laborers called the thin, half-putrid fish broth served every day 'graveyard soup' -- it contained nothing but bones.

"Prisoners with tiny children often looked on

helplessly as the toddlers sickened and died."

A former slave-labor camp inmate described how some prisoners acted after six months of constant hunger.

His story is told in Avraham Shifrin's 1982 *The First Guidebook to Prisons and Concentration Camps of the Soviet Union*: "You could hardly think about anything except a piece of bread.

"I witnessed a few prisoners who cut open a vein in order to dip bread in their blood as a means of satisfying their hunger.

"I once saw a prisoner who cooked a piece of flesh that he had cut from his own leg."

Terrible implements of torture are commonplace in Soviet slave labor camps.

Instruments are always readily available for ripping off the finger and toe nails!

There are special bone crushers and testicle smashers!

There are sole piercers for the victim's feet!

And there are a variety of electrical shockers!

Slave labor camp OTs-78/7 was located in Riga, the capital of Latvia.

45

Another former prisoner told Shifrin: *"Worst of all was RGB Captain Elmar Zaul, head of the operations section at the camp.*

"He was a real Gestapo type.

"In his office he had rubber hoses, boxing gloves, and even handcuffs attached to electrical wiring.

"Zaul thus often beat the prisoners himself!

"It was not unusual to hear the cries of his victims from his office."

"While the persecutions, arrests, massacres, mass starvations, and forced resettlements of people take place, those most capable of exposing such atrocities -- the Western governments, the media at best sit by quietly hoping not to embarrass the totalitarians," charged Vilius Brazenas who fled Communist Occupied Lithuania. *"When the nauseating stench of rotting corpses invades their nostrils; when their eyes witness the emaciated political prisoners; when they look about and discover that the totalitarians have destroyed the last vestiges of civilization, the people of the Free World will repeat the question others asked countless times before: 'How could we have allowed this to happen?'"*

 Americans should be aware of what to expect were Communists ever to achieve their avowed goal of world domination.

Eisenhower began to arm Communist Occupied Yugoslavia and Red Poland -- both miniature replicas of the Russian slave labor dictatorship!

Despite massive assistance Yugoslavia's Tito openly and repeatedly pledged his undying allegiance to Moscow: *"In peace as in war, Yugoslavia must march shoulder to shoulder with the Soviet Union."*

Stopping military aid to this bandit nation said Secretary of State John Foster Dulles (CFR) with a straight-face would force Yugoslavia back into the Russian camp!

This was claimed despite the fact that Tito was and always had been firmly in the Soviet camp!

The treasonous arms giveaways continued non-stop through subsequent administrations!

Even Ronald Reagan was afflicted with the malady referred to best as an infectious Militarily-Assist-The-Communist bug!

He somehow came to the paralogistic conclusions *"that furnishing of assistance to Yugoslavia is vital to the security of the United States.*

"That Yugoslavia is not controlled by the international Communist conspiracy.

"And that such assistance will further promote the independence of Yugoslavia from international Communism."

 During 1969 -- Nixon's first year in the White House -- over seven million pounds of America's strategic stockpile of tungsten was shipped to Communist Occupied Russia.

This metal is used for making armor-plating for tanks, assault helicopters, other military vehicles and armor-piercing shells.

48

Treason?

The Bryant Chucking Grinder Company of Springfield, Vermont first tried to export 164 Centalign-B ball-bearing grinders to Communist Occupied Russia in 1961.

The Reds claimed to need these highly sophisticated machines in order to build farm pump motors.

Senator Thomas Dodd didn't buy the phony pitch and intervened to stop the questionable deal!

President Nixon told C.L. Sulzberger of the *New York Times* on March 9, 1971: *"The Soviets now have three times the missile strength (ICBM) of ourselves.*

"By 1974, they will pass us in submarines carrying nuclear missiles."

Why then did his Commerce Department grant an export license in 1972 for shipping strategic ball-bearing grinders to Communist Occupied Russia?

Without Nixon's authorization the Centalign-B machines couldn't have been sold to America's deadly enemy.

The disastrous deal was consummated at the insistence of identified Red espionage agent Henry Kissinger.

Nixon's subversive shadow put all his prestige and influence behind the sale to his Kremlin masters.

These grinders produce precision micro-finish *(25 millionths of an inch)* ball-bearings which are used in the guidance systems of missiles and other space vehicles.

What did this advanced technology do for the criminal Russian dictatorship?

It allowed them to MIRV their missiles shortly after the sale!

In other words thanks to treasonous American assistance Communist Occupied

Russia could now fire missiles with multiple warheads.

Each warhead was to be programmed with a different target in the U.S.!

Respected consultant Dr. Miles M. Costick observed in 1976:  *"According to the intelligence estimates, by 1982, the Soviets will have at least 5,000 operational MIRVs aimed at the United States.*

"Without American computer technology and machines for production of precision miniature ball-bearings, this would not have been possible."

Said Dr. Susan L. M. Huck in October of 1976: *"This particular bit of treason, alone, should have cost everyone involved his freedom, if not his life. But of course it hasn't."*

Indeed decried Senator Jake Garn in November of 1981: *"Not only have Soviet ICBMs reached accuracies previously undreamed of by U.S. strategic analysts, but all Soviet military systems requiring precision inertial guidance have also reached a new level of accuracy and sophistication."*

51

This is a classic case of dual treason!

Treason involving the Nixon Administration and the Bryant Chucking Grinder Company!

Must our leadership be constantly reminded that terror has always been a major element in the Communist psyche?

Or do none of them really care?

Lenin was obsessed with the use of terror.

He spoke of it often, saying such things as *"terror must be encouraged."*

And *"we can achieve nothing unless we use terror."*

Moscow still honors Felix Dzerzhinsky who was the dreaded chief of the old secret police and a master terrorist.

Terror was a daily event when Communist Occupied Russia invaded Estonia, Latvia, and Lithuania.

People suspected of being anti-Russian were severely punished!

Nicolai Tolstoy tells how they were *"tied to trees, and there the guards experimented with various methods of prolonging death.*

"Some had their eyes slowly gouged out.

"Others were scalped and had their brains squeezed out of their skulls.

"Men had their tongues torn out!

"Their sides and legs slowly cut open!

"Or had bayonets slowly thrust into their mouths down their throats."

Where did the fiendish Red terrorists running Ethiopia get their tanks and rocket launchers and planes?

Guess?

Who gave Comrade Mengistu Haile Mariam the guns he needed to stay in power and commit unspeakable atrocities on his own people?

Guess who?

The United States under President Gerald Ford became Ethiopia's major weapons supplier in 1975 and 1976!

This Marxist dictatorship was given more military hardware than Ethiopia had ever dreamed of getting as a non-Red nation!

Boeing eagerly got in on the Help-The-Communists Gravy Train.

This company rushed to provide Communist Occupied Angola with new jets.

These aircraft were sold at terms extremely favorable to the People's Republic.

Boeing officials acknowledged they'd agreed to sell three 767 jet liners to Communist Occupied Poland's Lot airline in November 1988.

The cost: $220 million all of which as could be expected would be financed through Western banks!

Polish leaders refused to name the banks!

Why?

The answer should be self evident!

The world's most advanced seismological (earthquake detection) equipment was approved for sale to the USSR by the Commerce Department.

The Russians used this equipment to locate and track American missile-carrying subs!

Inedible tallow doesn't at first seem important.

The U.S. was regularly shipping tons of this to Communist Occupied Russia.

The Soviets used inedible tallow to manufacture TNT!

Senator Jake Garn gave other concrete examples of incredible American duplicity in June of 1986.

Some would call it outright treason: *"The Soviets obtained their shaped-charge technology used in war heads for antitank guided missiles from the U.S. oil-tool industry.*

"Since the early 1970s the Soviets and East Europeans have legally purchased more than 3,000 minicomputers, some of which are now being used in military related organizations."

Negotiations between Nicolae Ceausescu's oppressive Romanian slave state and United Aircraft were carried out for $100 million worth of heavy duty helicopters.

Lockheed made a lucrative business deal to build and deliver helicopters for Communist Occupied Russia.

Were these the devastating Mi-24 Hind/Assault helicopters the Soviets used in Nicaragua, Angola, Afghanistan, East Germany and elsewhere around the world?

Or are these Mi-24 flying tanks built for the USSR by some other treasonous American company?

The AFL-CIO Executive Council issued this belated warning: *"The list of recent Soviet acquisitions is enough to make Lenin's statement that the Western democracies would 'sell us the rope for their own hanging' seem like a prophecy."*

The AFL-CIO leadership stood virtually alone in its public acknowledgment that strategic war-related items could be utilized with *"nuclear warheads, armor-piercing rockets, and submarine detection devices."*

The Kama River truck factory was built in the Tartar Autonomous Soviet Socialist Republic.

This was the most spectacular example of America's treasonous assistance to Communist Occupied Russia to date.

It was designed by Americans, built by Americans and financed by Americans.

The plant was officially sanctioned by the Nixon, Ford and Carter Administrations.

It's located in Naberezhyne Chelny. `

This was a small village with no more than a few thousand population.

Where will the people come from to work in the factory and build the trucks?

Try some of the slave labor camps found on the Kama River!

Four of them are Chusovaya, Berezniki, Solikamsk and Vishersk.

In 1978 IBM sold Communist Occupied Russia computers worth $18 million for use in their Kama River truck plant.

This truck building factory was thus equipped with the world's largest industrial computer system!

This allowed the Reds to produce 250,000 diesel truck engines a year.

 The output of multi-axle 10 ton trucks at the Kama River plant surpassed that of all American truck builders combined!

The USSR dutifully promised to build only commercial vehicles.

Despite these promises the very first trucks to roll off the assembly line were used in 1979 to carry Russian troops into Afghanistan!

President Jimmy Carter said his eyes were rudely opened by the shocking Soviet invasion.

Yet IBM was allowed to keep a number of their computer experts working at the Kama River plant.

Here we find American citizens directly assisting in the production of Red Army vehicles!

Soviet troops were methodically decimating the population of Afghanistan in 1980.

Water supplies were poisoned and the earth was scorched in an effort to defeat the tough Afghan people.

Carter at this time treasonously ordered the Commerce Department to approve the sale to Communist Occupied Russia of automated assembly line equipment.

Why?

This high-tech machinery would double the output of the Kama River factory.

George P. Shultz (CFR) was later Reagan's Secretary of State and bore a great deal of responsibility for the existence of the Kama River truck plant.

So did Paul Volker (CFR) who was Chairman of the Federal Reserve Bank.

These men and their counterparts in other questionable deals with America's enemies were the ones Lenin referred to as *"deaf mute blind men."*

They loan money, give away technology and build the Communist military all the while laboring towards suicide.

Should not these and other American leaders have known better?

Certainly!

They were incredibly misinformed!

Incredibly naïve!

Or incredibly crass!

Or they deliberately committed treason!

It's certainly no secret that Red Army military vehicles have for decades been produced in American supplied Soviet auto and truck plants.

In 1980, the Dresser firm shipped to Communist Occupied Russia a $144 million factory to manufacture special petroleum drill bits for gas and oil exploration.

Located near Kuibyshev on the Volga River the completed plant was immediately converted to weapons production!

It produced armor-piercing shells specifically designed to stop American tanks.

This treasonous sale included a computer operated electron beam welder.

Such a high-tech welding machine is used in jet aircraft assembly. It also has a multitude of other military applications.

It's puzzling to observe *(or is it?)* American politicians and government bureaucrats continuing unabated to develop Communist Occupied Russia's military industrial base and equip their massive war machine.

These traitors or dupes know the brutal Moscow-based dictatorship has saturated our government, media, industry, religious organizations, military and other areas with massive spy networks.

They also know the atheistic tyrants entrenched in the Kremlin have never once deviated from their publicly declared goal of destroying the capitalist system and conquering the United States!

Some strategic items of distinct military application shipped to the Soviets have already proven disastrous.

For example Made in USA computers, radar and the missile firing guidance system was used by the Russian fighter pilot when the Kremlin ordered him to bring down Korean Airlines Flight 007 on September 1, 1983.

Communist Occupied Russia's most despised and feared anti-Communist was Congressman Larry McDonald.

He coincidently happened to be a passenger on that plane!

It's impossible for highly advanced computers to have no direct military applications.

Computer give-aways to the Soviet Union and other Red slave states are vital to their war-making potential!

Such treasonous dealings are a most serious breach of America's national security!

The Carter Administration allowed more than 300 new exemptions for exporting strategic goods to the Soviet Bloc.

Included was Control Data Corporation's Cyber 73 master computer.

The Cyber 73 was hidden in the Ural Mountains beneath four miles of granite.

From this location it directed a satellite network of IBM computers in place around Moscow.

They in turn operate Communist Occupied Russia's ABM (anti-ballistic missile) defense system!

Robert D. Schmidt was the Executive vice chairman of the Control Data Corporation.

63

He said this on April 1, 1986: *"We will never make our peace with right-wing individuals who oppose all trade with the Russians."*

Schmidt's firm sold over $50 million worth of computers to Communist Occupied Russia during the 1970s alone.

According to *Time* magazine of July 16, 1973: *"Control Data Corporation delivered the largest Western machine in the Soviet Union, a model 6200 now at the Dubna Research Institute."*

This computer was used for the development of missiles and other exotic weaponry.

FBI Director J. Edgar Hoover chided Schmidt and others like him in his speech to  the Daughters of the American Revolution (DAR) on April 22, 1954: *"To me, one of the most unbelievable and unexplainable phenomena in the fight on Communism is the manner in which otherwise respectable, seemingly intelligent persons, perhaps unknowingly, aid the*

Communist cause more effectively than the Communists themselves.

"The pseudo liberal can be more destructive than the known Communist because of the esteem which his cloak of respectability invites."

"Republicans pledge," said the Party platform of 1980, *"to stop the flow of technology to the Soviet Union that could contribute directly or indirectly to the growth of their military power."*

Furthermore the platform criticized the Democratic Administration of President Carter for allowing *"the most extensive raid on American technology by the Soviet Bloc since World War II."*

Apparently, these were no more than hypocritically hollow statements.

The record speaks for itself!

It all boils down to one simple inescapable fact.

The Executive branch of the government treasonously provides military hardware, weapons, equipment and training for brutal Communist police states!

These inhuman Red dictatorships have all undertaken large-scale executions of their own citizenry.

Each maintains dreadful concentration camps which supply unending numbers of slave laborers.

And they all without exception have a singular goal -- the destruction of Western civilization and especially the United States!

Despite all of this America continues to arm these tyrants and upgrade their weapons systems!

The genial appearing Mikhail Sergeyvich Gorbachev changed nothing.

 This ruthless dictator was simply Communist Occupied Russia's most cunning leader since Stalin.

He was no more than a chubby little Party functionary.

Moscow's version of the lovable great American showman, P.T. Barnum.

Gorbachev smiled, shook hands and kissed babies as if running for political office.

Yet the goals of his Communist regime remained the same.

The murder, rape, torture and naked Soviet aggression continued unabated!

"Transfers of strategic American technology to Communist Occupied Russia

66

are continually pushed by America's leaders," charged Anthony Sutton in his 1973 book, *National Suicide: Military Aid to the Soviet Union: "These men are fully aware that over 150 Soviet weapons systems are based on U.S. technology."*

And Sutton further stated: *"What we have done is given the Soviets the industrial and military capacity to destroy us".*

Solzhenitsyn warned: *"They are burying us alive and you are selling them the shovels."*

Because the United States arms the USSR the Russians in turn can afford to arm their surrogate slave dictatorships.

Communist Occupied Nicaragua is a prime example.

Defense Secretary Caspar Weinberger revealed that 23,000 metric tons of arms from the Soviets, East Germans and Cubans were shipped to the Communist Sandinistas in 1986.

The Pentagon said this included *"helicopter gunships as well as trucks and other vehicles, spare parts, communications equipment, armored vehicles, anti-aircraft guns, small arms, ammunition, and other military-associated cargo."*

"Years ago, when I was legal advisor to the [Soviet] Ministry of War Equipment, I  *spoke with many people who worked for the chief engineers and scientists,"* declared Avraham Shifrin, who spent 14 years in Soviet prisons for anti-Soviet activities. *"They openly told me that without American and European equipment, they absolutely could not work, because the calibrations and precision of Russian equipment were far too crude.*

"Their tanks are now standing on the threshold of Europe, their missiles in Cuba, pointed in the direction of the United States and their Embassies have proscription lists of the intelligentsia in your country marked for liquidation."

Dr. Edward Teller father of the hydrogen bomb was asked in September of 1983: *"If the Soviets launch a surprise nuclear attack against the United States, what would be the result?"*

His candid response: *"The question is when?*

"In a few years, if present trends continue, it is practically certain that it will be the end of the United States.

"The United States will not exist as a state, a power, an idea.

"It is possible that, in a few years, we shall be at the mercy of the Soviet Union, unless present trends change."

Ronald Reagan was asked by the *Los Angeles Times* for his view on the long heralded Soviet Red Chinese rift while campaigning for the Presidency in 1980.

He responded: *"They were allies and the only argument that caused their split was an argument over how best to destroy us."*

Asked if he would allow Communist Occupied China to obtain U.S. weapons: *"No, because they could turn right around and the day after tomorrow discover that they and the Soviets have more in common than they have with us."*

Mr. Reagan apparently had a *short* memory span.

Soon after his election America unhesitatingly jumped into bed with the world's bloodiest tyranny!

Why?

Reagan proceeded to dramatically increase the arming of Red China in an effort to boost the criminal regime to superpower level.

Why?

Americans were now propagandized with the same old song and dance routine!

We were advised that we must for reasons of national security and the advancement of world peace build the war-making capability of the despicable Red Chinese dictatorship.

Why?

The United States we were advised must develop Communist Occupied China into a first-rate military power.

Why?

We were given the same lame excuse as that used when U.S. leaders treasonously started militarizing the Soviets.

The Red Chinese had somehow suddenly become the *"good guys."*

They'll mellow say the *"experts"* but only if America modernizes its military.

In case of war with the Russian *"bad guys,"* our new Chinese pals will side with America.

Hogwash!

These self proclaimed *"experts"* must not have been listening to the late Chou En-lai.

He couldn't have made it any clearer in February of 1984: *"Some people may have thoughts of using Sino-Soviet differences to deal with China and the Soviet Union separately.*

"Those with such ideas will certainly be disappointed.

"On the contrary, if any act of aggression occurs against any Socialist country this would be an act of aggression against the whole Socialist camp.

"It would be impossible not to give support."

As Commander-in-Chief of America's Armed Forces, Reagan supplied the totalitarian camp with at least two  squadrons of American F-16 jet aircraft.

These planes were at that time among the most advanced fighter planes in the world.

 Keep in mind, Red China was the gangster nation that sent hundreds of thousands of fanatical *"volunteer" troops* to Korea when Americans were fighting and dying over there!

Close to 1,000 American POWs were known to be still held in North Korea and Red China, when the Korean War ended in 1953!

Many were shipped to Communist Occupied Russia's slave labor camps by the Korean Al Capones.

Red Chinese soldiers were also later sent into Vietnam to fight against and kill Americans in that no-win amphigory.

Export licenses were approved for shipping $500 million worth of military items to Red China in 1982.

President Reagan amended the nation's export-control law in 1983.

Communist Occupied China somehow overnight went from *"a nation hostile to the United States"* to a *"friendly, non-aligned country"* according to a story in the *Knoxville News Sentinel* on January 27, 1988.

It's no secret that Communist Occupied China runs more that 7,600 slave labor concentration camps!

They have an incredible 23 million slaves incarcerated in them!

This same inhumane Chinese dictatorship runs the world's largest single concentration camp.

Close to 10 million slave-labor prisoners are held in Chinghai, the northeastern province of Tibet.

Teng Hsiao-ping told the Third Session  of the Chinese Communist Party Central Committee in July of 1977: *"We belong to the Marxist Camp and can never be so thoughtless that we cannot distinguish friends from enemies.*

"Nixon, Ford, Carter, and future 'American imperialistic leaders' all fall in this category."

Nevertheless in 1983 the criminal giveaway of American military-related materials to the Chinese slave empire jumped to $740 million.

America's Chinese Advanced Military Technology Welfare Program was boosted to well over $1 billion in 1984.

The treasonous upgrading of Peking's antiquated military was enthusiastically endorsed by numerous elected and non-elected U.S. officials.

Senator Barry Goldwater acidly commented in April of 1986: *"These officials [might some be moles?] are so confident of the good faith and reliability of the Chinese Communist hierarchy and political system that they propose the United States Government should directly arm and modernize the military forces of a Marxist-Leninist regime founded on a one-party dictatorship."*

In September, 1983, Secretary of Defense Weinberger (CFR) visited Communist Occupied China.

He offered America's avowed enemies a new selection of advanced military hardware!

In March of 1984 Zhang Aiping, head of Red China's Ministry of National Defense, paid a secret return visit to the Pentagon.

Discussions centered around weapons purchases.

Included were anti-aircraft missile systems, amphibious tanks and huge 18-inch naval guns.

Tours of sensitive American defense plants were arranged.

Why?

So Aiping could do a little high-tech weapon window shopping!

James Gerstenzang wrote in the November 27, 1986 issue of the *Los Angeles Times*: *"The United States has been trying for several years to help nudge the China military toward the 1980s.*

"During his first visit to China as Defense Secretary in 1983, Weinberger initiated a military cooperation program.

"One of the major items resulting from his trip?

"Letting China use American technology to produce large-caliber artillery shells more cheaply than in the past."

"The United States," revealed *Insight* magazine on December 2, 1985 *"will deliver a $98 million ammunition plant capable of*

producing artillery munitions, explosives and other material. "

This was the first direct sale of military equipment and technology to Communist Occupied China by the U.S. government.

Included in the deal were the plans and materials necessary to construct a huge munitions plant.

This was done in an effort to improve the enemy's production time for making 155-millimeter shells.

Is this not treason?

Of course it was!

Reagan was somehow able to rationalize that selling weapons and other military hardware to Communist Occupied China was in America's best interest.

He said in a Presidential Order dated June 12, 1984: *"The furnishing of defense articles and services to the Government of China will strengthen the security of the United States."*

How?

How did America's newest pal and Red Chinese despot Teng Hsiao-ping feel about Reagan's statement?

He was quoted on October 13, 1986:

"Even though the American imperialists are the number one nation in scientific and technological matters in the future she will have no way of avoiding defeat by our hands."

Nevertheless, Aiping and other equally arrogant Communist Chinese thugs were again warmly welcomed to Washington in June of 1984.

This time Aiping was prepared to make massive purchases of military hardware with, of course, money borrowed at little or no interest from the United States!

American leaders groveled at the feet of the Reds as they went on a shopping expedition for weapons.

Export licenses for shipments to Communist Occupied China skyrocketed under the Reagan Administration.

They jumped from 2,020 in 1982 to an incredible 8,600 in 1985!

American exports to this corrupt slave labor dictatorship also exploded!

They jumped to $5.5 billion.

This was an increase of over 1,000 percent!

Yes this all took place while the Reagan Administration was in charge of everything!

Routinely approved for sale to the enemy were many high-tech military and military-related items.

This included torpedoes, radar for jet fighters, telecommunications gear, scientific instruments and equipment for manufacturing integrated circuits.

Restrictions on exporting strategic goods to Communist Occupied China were further relaxed on December 16, 1985.

Strategic items suddenly became *non-strategic* with the simple stroke of a pen!

America's most sophisticated computers, electronic instruments, precision machine tools and robotics could now be shipped to the Red Chinese with no restrictions.

Will Communist Occupied China leave Taiwan alone once they obtain enough U.S. military hardware to bring them into the Twentieth Century?

Not Hardly!

The Red Chinese leadership never even tried to mask their intentions!

Teng Hsiao-ping explained on October 13, 1986: *"Once normalization between China and the USA is finalized, it will naturally be beneficial to us in resolving the problem of liberating Taiwan."*

In other words, the Chinese Communist dictatorship fully intends to attack and conquer Taiwan as soon as they feel the time is right!

How did the President react?

Incredibly, Reagan responded with a clear unmistakable invitation for this bandit regime to move against Taiwan: *"The problem between the People's Republic and the people of Taiwan is one for the Chinese to settle between themselves.*

"We will do nothing to intervene."

Communist Occupied China's plan for *"liberating"* Taiwan will follow a carefully scripted scenario.

Their horrifying subjugation of Tibet gives an accurate picture of what may be expected.

Offers James J. Drummey: *"In their effort to exterminate the Tibetan race, culture, and religion, the Reds killed scores of*

thousands of people, destroyed more than 2,000 monasteries, imprisoned countless numbers of Buddhist priests and nuns, shipped children to Red China for indoctrination in Communist ideology, and virtually wiped the nation out of existence."

Yet the President of the United States callously tells them: *"We will do nothing to intervene!"*

Admiral James D. Watkins Chief of Naval Operations led a special delegation to Communist Occupied China in April 1986.

His assigned task of treason was to

work out the details of yet another massive weapons and technology sale to America's Red enemies.

Included in this particular giveaway package were sonar devices, anti-submarine warfare helicopters and turbine engines worth hundreds of millions of dollars.

May of 1986 saw the arrival in Washington of the notorious General Yang Dezhai, Chief of Staff for the People's Liberation Army.

This war criminal was treated like royalty although he had personally led drug-crazed Chinese troops against American boys fighting in Korea.

Despite his well known role in the Korean War this Twentieth Century savage was actually wined and dined in the White House!

Many U.S. leaders breathlessly shook Dezhai's hand, stained though it was with the blood of American soldiers and marines.

General Dezhai had private meetings with Defense Secretary Weinberger (CFR); Vice President George Bush (CFR); and Chairman of the Joint Chiefs of Staff Admiral William Crowe (CFR).

He had the audacity to ask that the U.S. participate with Red China in joint military exercises.

What exactly did this bedlamite have in mind?

To again use American servicemen for target practice as he did in Korea?

This was the horrifying dictatorship the Reagan Administration sold hi-tech torpedoes to in 1986.

Defense Secretary Weinberger said to the *Los Angeles Times* on November 27, 1986, that he was *"making available to the Chinese navy an upgraded Mark 46 torpedo to improve the Chinese anti-submarine capability."*

Red Chinese sailors were to be trained in torpedo maintenance and use!

Where?

At the U.S. Naval Training Center in Orlando, Florida.

The sale of advanced avionics equipment to Communist Occupied China was announced to Congress by the Reagan Administration on April 8, 1986.

These high-tech military items were innocuously described as *"55 integrated avionics systems kits, support equipment, training, and system installation."*

They were to be installed on Red China's F-8 jet fighters.

Included were airborne radar units, navigation gear, special computers and targeting panels.

Secretary of Defense Weinberger said this in the *Los Angeles Times* on November 27, 1986 *"the improvement of the F-8s through the sale of $550 million worth of electronics and radar systems is considered particularly important."*

Why?

The contractors selected to manufacture these high-tech items for the Communist Chinese were also to loan the Reds as many as 25 of their top engineers.

These Americans would install the equipment and train their Chinese counterparts in its operation and maintenance.

Lastly, five members of the United States Air Force would be loaned to Red China for up to six years!

For what?

They'd have the unsavory task of training Chinese pilots.

It's well worth remembering that some of the American flyers downed over Vietnam -- members of the same United States Air Force -- were still being held as POWs in barbaric Red Chinese prison camps!

A nameless faceless bureaucrat deeply buried in the bowels of the Defense Department responded with a straight-face: *"This sale will contribute to the national security of the United States by helping to improve the security of a friendly country which has been an important force for political stability and economic progress in Asia and the world [as in North Korea, Vietnam, Laos, and Cambodia?].*"

Politics being as they are another unnamed obscure bureaucrat in the Defense Department at the same time classified Communist Occupied China as *"hostile to the United States."*

"We are at war," said Ronald Reagan during the 1964 Presidential Campaign of Senator Barry Goldwater, *"with the most dangerous enemy that has ever faced mankind."*

Then exactly what was his thinking behind letting Red Chinese soldiers take airborne training at Fort Benning, Georgia?

Why then did this President sponsor the sale of a $98 million ammunition plant to Communist Occupied China?

Did Mr. Reagan have a change of mind on the way to or upon becoming the President of the United States?

Or had he simply been acting all along?

Epilogue

The record covering crucial episodes of the McCarthy era has been massively and deliberately distorted from the very beginning!

Conveniently forgotten or deliberately overlooked are the 78 hearings held between 1951 and 1952 by Senator William E. Jenner's (R-Indiana) Senate Internal Security Subcommittee (SISS); the House Committee On Internal Security; the House Un-American Activities Committee (HUAC) under the chairmanship of both Martin Dies (D-Texas) and Francis Walters (D-Pa); the Federal Bureau of Investigation (FBI) under the guidance of J. Edgar Hoover; and other investigating committees and individuals.

Out of all of these investigations one man was selected:

To be stopped!

To be destroyed!

To be made an example!

Why?

So that no one would ever again dare to initiate any investigations into the penetration of our government agencies by communist agents (spies) in the employ of the Soviet Union!

Yes!

An obscure Senator from Wisconsin was deliberately targeted for this purpose!

Joseph McCarthy's incredibly successful investigations panicked those on the political left.

Their reaction was shockingly quick!

Key data was been suppressed, denied and even widely falsified.

This took place in the media, all branches of government and many alleged scholars entrenched in the ivory towers of our institutions of higher learning!

Such misreporting and misrepresentation of the facts continues today.

Much of the misinformation we were (and still are today) so carefully spoon-fed about Senator Joseph McCarthy the man and his investigations was no more than an admixture of uncheckable blovations from deceased third parties and demonstratable falsehoods!

For example, how many innocent people were harmed by McCarthy's revelations?

The correct answer?

Not one!

No!

Not One!

McCarthy's most virulent critics have had more than a half century to produce the names of the hundreds of innocent people they claim were destroyed by the astounding revelations of the Senator from Wisconsin.

Yet those highly skilled propagandists in our media and government and institutions of higher learning have been unable to name even one innocent person they claim was destroyed after being falsely accused by McCarthy!

How many innocent people committed suicide as a result of McCarthy's exposure?

The correct answer?

Not one!

Not one suicide can be attributed to the investigations conducted by McCarthy!

No! Not one!

According to the obscene claims made the highly skilled propagandists in our media, government and scholars entranced in those ivory towers of our colleges and universities there were a rash of suicides with bodies falling constantly of the heads of pedestrians below on the streets of Manhattan!

Once again, McCarthy's most virulent critics have had more than 50 years to produce the names of the hundreds of innocent people they claim committed suicide because of the astounding revelations of the Senator from Wisconsin.

Yet those highly skilled propagandists in our media and government and institutions of higher learning have been unable to name even one innocent person they claim committed suicide after being falsely accused by McCarthy!

No!

Not one!

But there were two suicides on record during the McCarthy period!

Neither was the result of an innocent person who'd been ruined by McCarthy's revelations!

Both were subversives who'd been exposed by McCarthy!

Both were subversives who'd been positively indentified as Kremlin agents!

Lawrence Duggan had been operating in the State Department as a widely known Soviet spy!

He'd been called to testify before a Congressional investigating committee.

Duggan never made it!

He conveniently "fell" from a window high up in a Manhattan skyscraper!

Fell?

Probably not!

He was more than likely pushed from or tossed out of the window by an assassin in the employ of the Soviet Union!

Why?

To make certain he didn't fold under pressure and start naming other Kremlin moles.

Secondly there was the unexpected demise of Harry Dexter White.

This Soviet agent discovered that he was being investigated by J. Edgar Hoover of the FBI!

He died of a sudden heart attack!

Coincidence?

Not hardly!

Was White's death a suicide?

Yes or at least so claimed McCarthy's critics!

Again, not hardly!

Heart attacks can readily be induced with the proper use of certain medicines administered by a hired assassin in the employ of the Kremlin!

Why?

Simply to eliminate anyone who might panic and decide to turncoat and reveal the names of other spies secretly entrenched deeply in the bowels of every branch of our government.

To sum up, most fit into one of three categories:

Conscience lacking incurable liars!

Those with an axe to grind!

Individuals who simply do not know the facts!

If you liked this book in the *None Dare Call It Treason* series then you'll probably also enjoy reading the others!

Gift copies of this book can be ordered at

createspace.com/4216634

Available Titles

None Dare Call It Treason Book 1
The Internal Security Farce!
5.5" x 8.5" 97 pages $4.95
Order from createspace.com/4215951

None Dare Call It Treason Book 2
Never Ending Subversion
In Government!
5.5" x 8.5" 202 pages $4.95
Order from createspace.com/4216385

None Dare Call It Treason Book 3
America's Subversive State Department
Bloated With Security Risks
5.5" x 8.5" 202 pages $4.95
Order from createspace.com/4216626

None Dare Call It Treason Book 4
America's Illustrious State Department!
It's Machiavellian Misdeeds!
5.5" x 8.5" 202 pages $4.95
Order from createspace.com/4215018

None Dare Call It Treason Book 5
Our Presidents A Major Security Threat!
5.5" x 8.5" 202 pages $4.95
Order from createspace.com/4213501

None Dare Call It Treason Book 6
Presidential Words & Deeds
&Blatant Lies!
5.5" x 8.5" 202 pages $4.95
Order from createspace.com/4213920

None Dare Call It Treason Book 7
Subversives Close To Our Presidents
5.5" x 8.5" 89 pages $4.95
Order from createspace.com/4213931

None Dare Call It Treason Book 8
Henry Kissinger
The Shadowy Untouchable Kremlin Spy!
5.5" x 8.5" 202 pages $4.95

Order from createspace.com/4214986

None Dare Call It Treason Book 9
Inexcusably Arming America's Enemies!
5.5" x 8.5" 202 pages $4.95
Order from createspace.com/4216634

None Dare Call It Treason Book 10
*Inexcusably Financing
America's Enemies!*
5.5" x 8.5" 202 pages $4.95
Order from createspace.com/4216777

None Dare Call It Treason Book 11
*Treasonous Trade With & Aid To
Enemies Of Freedom!*
5.5" x 8.5" 202 pages $4.95
Order from createspace.com/4216873

None Dare Call It Treason Book 12
*Wholesale Treason During the War
In Vietnam!*
5.5" x 8.5" 202 pages $4.95
Order from createspace.com/4215293

None Dare Call It Treason Book 13

Big Business
& Astounding Acts Of Treason!
5.5" x 8.5" 202 pages $4.95
Order from createspace.com/4215805

None Dare Call It Treason Book 14
Illegally Importing
Slave Made Goodies!
5.5" x 8.5" 202 pages $4.95
Order from createspace.com/4215894

None Dare Call It Treason Book 15
The House That Hiss Built
The Anti-American United Nations!
5.5" x 8.5" 202 pages $4.95
Order from createspace.com/4215323

None Dare Call It Treason Book 16
Security Risks in the House and Senate!
5.5" x 8.5" 202 pages $4.95
Order from createspace.com/4213508

None Dare Call It Treason Book 17
The Supreme Court A Devastating
Threat To National Security!

5.5" x 8.5" 202 pages $4.95
Order from createspace.com/4213689

Orders for Resale
40% Off Retail Price

Send Purchase Order to

christianamerica2@yahoo.com

MEET THE
AUTHOR

Robert W. Pelton has been writing and lecturing for more than 45 years on political, religious and historical subjects.
He has published more than 100 books including the sensational exposé *Unwanted Dead or Alive – The Greatest Act of Treason in Our History – The betrayal of American POWs Following World War II, Korea and Vietnam.*

Mr, Pelton proudly claims a heritage going all the way back to well before the War for American Independence.

One of Mr. Pelton's ancestors, John Rogers, came to America on the Mayflower and was one of 41 signers of the Mayflower Compact.

Another, John Smith was one of the founders of Jamestown.

Peleg Pelton served as the fifer in the Continental Army at age 18 during the Battle of Saratoga (1777) and again in Yorktown (1781).

Captain Peter Hager was Commander of the Old Stone Fort in Schoharie, New York, in 1780.

Another, Captain Bezaleel Tyler fought in the only Revolutionary War Battle taking place in Sullivan County, New York.

Mr. Pelton is a member of Sons of the Revolution (SOR), and Sons of the American Revolution (SAR).

www.ingramcontent.com/pod-product-compliance
Lightning Source LLC
Chambersburg PA
CBHW071721170526
45165CB00005B/2108